Jorge Alfredo González Pérez.
Natalia Patricia Fatima Do Carmo

BELL PEPPER CROP RESPONSE TO DIFFERENT PLANTING DENSITIES

Jorge Alfredo González Pérez.
Natalia Patricia Fatima Do Carmo

BELL PEPPER CROP RESPONSE TO DIFFERENT PLANTING DENSITIES

RESPONSE OF THE BELL PEPPER CROP TO DIFFERENT PLANTING DENSITIES IN THE SOIL AND CLIMATE CONDITIONS OF ONDJIVA

ScienciaScripts

Imprint
Any brand names and product names mentioned in this book are subject to trademark, brand or patent protection and are trademarks or registered trademarks of their respective holders. The use of brand names, product names, common names, trade names, product descriptions etc. even without a particular marking in this work is in no way to be construed to mean that such names may be regarded as unrestricted in respect of trademark and brand protection legislation and could thus be used by anyone.

Cover image: www.ingimage.com

This book is a translation from the original published under ISBN 978-3-639-69038-5.

Publisher:
Sciencia Scripts
is a trademark of
Dodo Books Indian Ocean Ltd. and OmniScriptum S.R.L publishing group

120 High Road, East Finchley, London, N2 9ED, United Kingdom
Str. Armeneasca 28/1, office 1, Chisinau MD-2012, Republic of Moldova, Europe
Managing Directors: Ieva Konstantinova, Victoria Ursu
info@omniscriptum.com

Printed at: see last page
ISBN: 978-620-8-56770-5

RESPONSE OF THE PEPPER CROP (CAPSICUM ANNUUM) TO DIFFERENT PLANTING DENSITIES IN SOIL AND CLIMATIC CONDITIONS OF ONDJIVA.

AUTHOR(S): MSC. JORGE ALFREDO GONZÁLEZ PÉREZ

ENG NATALIA PATRICIA FATIMA DO CARMO

CUNENE, 2024

ACKNOWLEDGEMENTS

✓ *To God for life and for helping me to overcome all the obstacles encountered throughout the course,*

✓ *To my parents and siblings, who encouraged me in difficult times and understood my absence while I dedicated myself to doing this work,*

✓ *I mustn't forget the group of teachers for their corrections and teachings, which have enabled me to perform better in my professional training process, especially Professor Jorge Alfredo González Pérez, who is largely responsible for the realisation of this work,*

Thank you all very much, God bless you richly and powerfully.

DEDICATORY

I dedicate this work to the special and important people in my life:

- ✓ *My mother, Mrs Maria Domingos Fatima (in memory);*
- ✓ *To my father Jose Docarmo Mateus;*
- ✓ *To my children for being the reason I face every battle and the reason for my life;*
- ✓ *To my brothers who have always supported me on this long journey;*
- ✓ *To my husband Gilson Vahekeny, who has always been by my side to make this dream possible.*

SUMMARY

This research was carried out on João Kambanda's property located in Oshomokuyo in Ondjiva Province. The aim of the research was to evaluate the effect of different planting distances on the cultivation of peppers (Capsicum annuum) of the California Wonder variety, studying the morphophysiological and agricultural parameters of the plant in open field conditions. The experimental design used was a completely randomised block design with four treatments and three replications. Morphophysiological parameters and crop yield were assessed, and the results obtained were statistically analysed for comparison. The best results obtained in the open field in terms of morphophysiological parameters was treatment T1 with a planting distance of 0.40 m x 0.20 m with an average height of 36.03 cm, a stem diameter of 0.62 cm and 2 fruits per plant. The physiological indices also corresponded with a planting distance of 0.40 m x 0.20 m, a leaf area index of 0.11 m^2, a photosynthetic potential of 2.82 m^2/day, a TON of 0.50 g.m^2.day-1 and a crop yield of 4.82 kg/$m^{(2)}$.

Key words: morpho-physiological, treatment

CONTENTS

ACKNOWLEDGEMENTS 2

DEDICATORY 3

SUMMARY 4

INTRODUCTION 6

LITERATURE REVIEW 10

MATERIALS AND METHODS 19

RESULTS AND DISCOUNT 27

CONCLUSIONS 39

RECOMMENDATIONS 40

BIBLIOGRAPHY 41

ANNEXES 47

INTRODUCTION

Peppers (Capsicum annuum) are a species native to Mexico, Bolivia and Peru which, over the years, has adapted to cultivation in various countries around the world. This vegetable is one of the most sought-after by consumers when it comes to incorporating them their diet (FAO, 2010).

Árias (2013) explains that peppers are one of the most in-demand crops worldwide and nationally due to their high consumption. This product is used in different forms: fresh peppers, powdered peppers for condiments such as paprika and canned peppers. There is a high demand in the European markets for fresh peppers all year round, and their production has increased on the Spanish Mediterranean coast as new techniques have been implemented to produce peppers in the open, under cover or in greenhouses, as well as applying different sowing densities in these systems

Sowing density is one of the main factors influencing the production, quality and, in some cases, the earliness of crops (Agarwal et al., 2007). Because sweet chile has a dichotomous growth, a high sowing density can have an effect on plant development; however, this variation can be manipulated by changes in the structure of the plantation (Cerque Regeu et al., 2009).

Worldwide, the cultivation of vegetables is an important activity because of the benefits it has for human nutrition and among these, peppers stand out, as they are a fruit with a high vitamin C content, are rich in calcium and phosphorus and also have a high level of fibre, which highlights their benefits for a healthy diet.

Borbor and Suárez (2007) state that in Ecuador pepper production is an important figure in the agricultural sector, and that it is a species that is grown both on the coast and in the inter-Andean valleys. According to the last National Agricultural Census (2000), approximately 956 hectares were cultivated as a monoculture and 189 hectares as an associated crop, with the main coastal provinces of Guayas, Manabí and Esmeraldas producing the most. The country began exporting in 1996, with Spain and the Netherlands being the main markets; in order to increase yields per hectare of this vegetable, pepper producers are using new hybrids.

Nogueira et al. (2007) emphasise that in order to meet the demands of the horticultural crops market, it is necessary to introduce new ways of managing production systems, technologies and varieties that adapt to local climatic conditions. Studies carried outunder controlled conditions (greenhouse) for peppers point out that it is necessary to analyse factors such as humidity, temperature, light, pests, diseases, fertilisation, sowing density, among others.

Often, the low yields achieved by farmers are due to the fact that can't control one of the climate factors, which in recent years has been one of the biggest constraints on the production process for farmers.

Nowadays, many growers have opted to produce new hybrids to increase yields per hectare. Rios (2012) states that the California Wonder pepper variety has a cycle of 90 days after transplanting, is fast-growing, has a wide fruit, ends in a point, has no shoulders, is green to red when ripe, and weighs between 170 and 220 g, with a yield of 140,052.00 kg/ha.

The crop's rusticity allows it to be grown in unfavourable conditions, as well as in very cold temperatures, with excellent results. The advantages of sowing the California Wonder variety are as follows: the shoulderless fruit does not accumulate rainwater, so you don't get rotten fruit, it has excellent colour and flavour, which are key characteristics for greater consumer acceptability.

Martínez (2005) states that different planting distances are used: 1.20 m to 1.80 m wide and double rows 0.30 cm between plants, 0.30 and 0.45 cm between rows, so that there is no competition for light, water or contact between the fruit and the soil, and the planting frame provides increased aeration for the plants.

Alemão et al., (2014) states that in the African region, agricultural and biological yields for growing peppers under glass and in the open field are low, due to the influence of abiotic factors in the region, based on the fact that the climatic conditions and soils of the region are largely unsuitable for this crop due to its vegetative development.

Optimum planting distances for both production systems are not yet available, especially in the open field. The information available from these studies on the yields and morphological behaviour of the variety (California Wonder) is a limiting factor when it comes to making decisions to improve production technologies and yields for this vegetable in the African region. Because of the above, we have limited experience in producing peppers at different planting distances suitable for this crop. In the different gardens that we wanted to analyse from a uniform planting distance in the Ondjiva locality, we couldn't see a planting landmark as uniform as possible that we could take as a starting point or witness for our investigation.

This is why we emphasise that the **basic problem of** this research is based on the following:

How can the morphophysiological development and agricultural yield of the pepper crop (Capsicum annuum) be increased under different planting distances in the soil and climatic conditions of Ondjiva province?

We have formulated the following objectives for this problem:

General objective: "To evaluate the effect of different planting distances on the cultivation of peppers (Capsicum annuum), on morphophysiological and agricultural parameters under soil and climatic conditions in the Odjiva province."

The following **specific objectives** are formulated for this general objective:

a) To characterise morphophysiological indicators and agricultural yield in the cultivation peppers (Capsicum annuum) at different planting densities

b) To determine the most suitable population density for growing peppers

(Capsicum annuum) in the soil and climate conditions of Cunene Province.

Based on the following **hypothesis**: "If a planting density for pepper cultivation is determined for the soil and climatic conditions of Cunene Province, then yield increases would be achieved in pepper cultivation (Capsicum annuum).

LITERATURE REVIEW

2.1. Pimento Generalities

Collantes (2015) says that peppers are a small annual shrub between 0.75 and 1.0 m tall, belonging to the Solanaceae family, with a fragile, erect, green stem, with branches that split into two parts. Its leaves are large, bright deep green colour, oblong (wider than wide), lanceolate or globose. Its flowers are scarce and white or yellowish-white in colour. Its sowing density is approximately 30,000 plants per hectare and it is propagated by seeds.

Harvesting begins between 90 and 115 days after sowing or planting and its vegetative life cycle lasts two or three months after the first harvest. It adapts well to hot climates and does not tolerate freezing temperatures. It needs 1000 mm of rainfall and is a short-day silver that needs a temperature of between 21 and 26 °C, but it should not go below 16 °C.

Alemão et al, (2017) report that in the open field, pepper plants were 63 cm tall at 70 days after transplanting, while under glass the plants were 136.3 cm taller, with significant differences between the treatments. This variable is a function of luminosity, if it is low the internodes of pepper stems widen, resulting in greater plant height and taking into account that under glass you can control factors such as humidity, temperature, luminosity, pests, diseases, fertilisation, sowing density, among others.

2.2. Pepper taxonomy

According to Rios (2011), the botanical classification of peppers is summarised follows:

- Kingdom: Vegetable,
- Sub-kingdom: Embribionta,
- Division: Magnoliophyta,
- Sub - Division: Magnoliopsida,

- Class: Asteridae,
- Order: Solanales,
- Family: Solanaceae,
- Genus: Capsicum,
- Species: Annuum.

2.2.1. Radical system

Rios (2012) states that the pepper plant has a pivotal and deep root (depending on the depth and texture of the soil), with numerous adventitious roots that develop horizontally and can reach a length of between 50 cm and 1.0 m.

2.2.2. Main stem

Pérez (2014) states that the stem of the pepper is erect, reaching a height of 30 to 45 cm from a certain height ("cross"), emits 2 or 3 branches depending on the variety and continues branching dichotomously until the end of its cycle, with a stem diameter of 1 to 2.02 cm, the secondary stems fork after emitting several leaves, and so on.

Pérez et al., (2011) reported that pepper plants obtained a stem diameter a row of 1.5 to 1.7 cm, which allows the plant to have good branch and fruit support. Alemão et al. (2017) reported that the values obtained for stem thickness under glass ranged from 0.54 to 0.57 cm. This variable is an indicator of the vigour and strength of California Wonder pepper plants.

2.2.3. Leaf

Orellana and León (2011) describe that the pepper leaf is entire, imbibed and lanceolate, with a very pronounced acuminate apex and a long, inconspicuous petiole. The bundle is glabrous (smooth and soft to the touch), more or less intense green in colour depending on the variety and shiny.

The main vein starts at the base of the leaf, like an extension of the petiole, as do the secondary veins, which are pronounced and reach almost to the edge of the leaf. The leaves are inserted alternately into the stem and vary in size depending on the variety, with a certain correlation between the size of the adult leaf and the average weight of the fruit.

Alemão et al. (2017) reported that leaf production under glass and in the open field corresponded to 50.3 and 80.3 leaves, with significant differences at 72 days after transplanting. This is an indicator to consider for pepper cultivation, as they are varieties with indeterminable growth and leaves are a fundamental organ for the photosynthetic process of this crop.

2.2.4. Flower

Morais and Pachacama (2011) state that pepper flowers appear solitary and are made up of whitish flowers located in the leaf axils, with one flower per node. They are made up of five united petals and five independent petals.

2.2.5. Fruit

Cartagena (2004) describes that the pepper fruit is a hollow, semi-cartilaginous, depressed berry of variable colour (white, red, yellow, orange, violet or green); some varieties go from green to orange and red as they ripen. It varies in size, weighing from a few grams to over 500g. The seeds are inserted into a centrally arranged conical placenta. They are rounded, slightly reniform, pale yellow in colour and vary in length between 3 and 5 mm.

Reis et al. (2017) reported that the number of fruits per pepper plant ranged from 33.67 to 14.66, which may be due to the factors that the plants require for their development. Pérez et al. (2011) reported that the number of fruits per plant ranged from 7.2 to 11.8, taking into account that this experiment studied non-hybrid varieties of peppers.

2.3. Non-hybrid pepper variety California Wonder

Logistics and industrial centre Qbox-Bod. 25, Copacabana, Antioquia (Colombia) (2023), the pepper is a vegetable with the scientific name AAPSICUM ANNUUM and the common name pimentón, it is a variety originating from Sudamerica and proliferates at an altitude of 0-2000 metres above sea level, it is a plant that can reach between 40 and 50 cm, It has a sowing density per hectare of 1 to 3 lb, provides 140 to 160 seeds per gram (gr), has a harvest cycle of 90 days and can reach a production of 15 to 20 $tonnes.ha^{-1}$. In general, the fruit is produced at different times and varies in shape and size. They are usually conical, obtuse-pointed, smooth on the surface, first green and then purple, 7 to 10 centimetres long and 6 to 9 centimetres long, with thick ends. The best soils are sandy loams, deep, rich in organic matter and mainly well-drained. The optimum pH ranges from 6.5-7, which can withstand certain acidic conditions.

Origin

Peppers are a sweet species of the Capsicum genus, which is part of Solanaceae botanical family. It is the most widely cultivated species of this genus. They all originated on the American continent, probably in what I understand to be the southern part of Brazil, but Colombia is also considered to be one of the centres of origin (Corporación Colombiana de Investigación Agropecuaria (Corpoica, 2014)).

2.4 Agro-climatic characteristics for growing peppers (Capsicum annuum)

Diho (2008) states that low temperatures do not favour the development of the flower bud, temperatures between 10 °C and 15 °C cause the flowers to deform and fall off with any of the following anomalies: curved and undeveloped petals, formation of multiple ovaries that can develop into fruits distributed around the main one, shortening of the wool threads and pistils, thickening of the ovary and pistil, fusion of anthers, etc.

They also lead to the formation of smaller fruits, which can deformed, reduce pollen viability and favour the formation of parthenocarpic fruits. The optimum relative humidity is between 50 and 70 per cent. If the humidity is too high, it leads to the development of diseases in the parts of the plant and makes fertilisation more difficult. If, on the other hand, the humidity is too low, during the summer, with high temperatures, the plants abort the flowers and freshly curdled fruit.

It is only after the first fruits have set that the plant tends to balance its vegetation and fructification. In addition, peppers are cool-climate plants that can withstand temperatures between 18 °C and 30 °C, which means that in the tropics they should be sown in the winter months or in high places (Moreno et al., 2004). If it is grown at temperatures above 35°C, fructification is poor and the fruit is burnt or scalded by the action of the sun's rays.

Berrios, Arredondo, Tjalling, (2007) state that plants absorb radiation in their chlorophyll cells of a wavelength ranging from 400 - 700 NM and use it as energy for photosynthesis to transform CO2 into sugar. This radiation is called RAF (Photosynthetic Active Radiation), expressed in $J.m^{-1}.s^{-1}$ RAF determines the amount of sugar produced in the leaves during photosynthesis.

The higher the amount of sugars produced, the higher the fruit load the plant can withstand and the higher the yield. RAF is responsible for 45 - 50 % of global radiation (300 - 1100 NM). Many computerised control systems in greenhouses use radiation measurements.

2.5 Soils

Domínguez (2009) states that the most suitable soils for growing peppers are loamy - sandy, deep, rich, containing a proportion of organic matter of 3 to 5 per cent.

- 4 % and well-drained soils.

For its part, the optimum pH values range from 6.5 to 7, although it can withstand certain acidic conditions (up to a pH of 5.5); in sandy soils it can be grown with pH values close to 8. As for irrigation water, the optimum pH is between 5.5 and 7.

Preparing the soil before sowing is one of the most important agricultural practices. Proper soil preparation will facilitate the optimum growth and development of the plant's roots, which will facilitate the extraction of water and nutrients from the soil. This practice incorporates existing plant residues, improves soil aeration and drainage, facilitates the decomposition of organic matter and favours the control of soil pests and diseases (Martínez, 2005). AGRIPAC (2012) states that soil conditions should be reviewed for optimum plant development.

2.6 Sowing, planting and cultivation of peppers

Seed propagation is recommended for pepper production. For one hectare you need 600 to 700 grams of seed, which should have a germination rate of no less than 83 per cent. It is important to take the following considerations into account when preparing a good seedbed: to sow the seed, soak it for 18 to 24 hours in running water, then in hot water for 4 to 8 hours or in milk for 12 hours. In addition, the optimum germination temperature is between 18ºC and 35ºC. The seed should be disinfected with Vitavax® at 100 - 200 g/100 kg of seed to avoid disease problems later on (AGRIPAC, 2000).

Plants should be transplanted 30-35 days after germination when they are at least 15 cm tall and have 4-5 true leaves. It is advisable to do this in the early hours of the morning or in the cooler hours of the afternoon to reduce stress on the plants and thus obtain good survival results. In addition, a root disinfectant (Captam and Terraclor) should be applied before transplanting (Villavicencio and Vásquez, 2008).

Eco Agriculturist (2014) reports that the planting distance depends on the size of the plant, which in turn depends on the commercial variety grown. The distance most often used in greenhouses is 1 m between rows and 0.5 m between plants, although when it comes to medium-sized plants and depending on the type of pruning, it is possible to increase the planting density to 2.5 - 3 plants per m^2. It is also common to arrange rows of crops in pairs, 0.80 m apart and leave aisles of 1.2 m between each pair of rows in order to favour cultivation work and avoid undesirable damage to the crop. Under glass, the planting density is usually between 20,000 and 25,000 plants/ha.

Outdoors, they are used to growing up to 60,000 plants/ha. Formative pruning is a normal cultural practice in greenhouse conditions. There are different forms of pruning that have the effect of producing fruit that is of higher quality for commercialisation. Pruning vegetables allows the plants to be kept vigorous and ventilated, so that the fruit is not hidden by the foliage and is safe from the sun (Agropecuarios, 2014).

The normal number of stems used grow the vegetable is 2 or 3 branches. If necessary, leaves and shoots located under the "cross" should be cleaned (Eco Agriculturist, 2014).

He also points out that the plant should be supported so that it stays upright, as the pepper stems break very easily. It is important to remove the weeds during the pepper's growing cycle in open field conditions and under . Weeds should be controlled by three or four weeding operations, using small hand tools such as azadillas, binadoras, scarifiers or mechanical weeders that come with sharp, hardened steel handles.

Stripping must be carried out very carefully to avoid causing the roots to move and damaging the plant's root system (Suquilanda, 2002).

The same author states that fruit lightening is recommended (eliminating the fruit that forms on the first "cross") in order to obtain larger, more uniform and earlier fruit and higher yields. In plants with little vigour or hardened by cold, high salinity or unfavourable environmental conditions, very small and poor quality fruits are produced, which should be eliminated by lightening.

2.7. The global importance of peppers

Peppers (Capsicum annuum) are a very profitable crop and nowadays there is some research that indicates that vegetables can be produced under glass and in the open field.

Moreno et al., (2011) reported that the pepper plants reached a maximum height of 80.9 cm and a minimum of 50 cm, these differences being due to the length of the internodes and not the number of internodes, they obtained an average of 11.8 fruits per internode. plant, its average weight was 178.7 g, it was 9.2 cm long and 8.5 cm long. According to Cabalceta et al. (2017), the production of 15 genotypes under glass showed a height growth row between 0.82 m and 1.44 m, this variable is related to the number of nodes or the length of their internodes, in the same experiment a leaf area of 1.22 m^2 was obtained under glass.

Pérez et al., (2017) show that in the production of peppers (Capsicumm annuum L.), in the open field, the best indicators were stem thickness 0.6 to 1.3 cm, in addition to crop yield parameters, 9 fruits were obtained per plant, with a total fruit weight of 975.80 g, and a leaf area of 0.95 m[2.].

Rodríguez et al. (2014) state that peppers produce two or three branches, depending on the variety, and continue branching dichotomously until the end of their cycle, with the secondary stems forking after several leaves have sprouted, and so on. Yields ranged from 1.15 kg/plant to 0.77 kg/plant. The same author states that the potential yield of peppers in the open field ranged from 11 to 13 tonnes per hectare $^{-1}$, and the weight of the fruit obtained in the research was approximately 232 grams.

According to Reis et al. (2017), the number of per pepper plant fluctuated between 33.67 and 14.66, which can be attributed to factors that plants require for their development. The diameter of the fruit with the highest result for this research at the first harvest was 11.78 cm, with no significant differences from other treatments.

Alemão et al., (2017) show that the biological yields under glass were higher than those in the open field with 46.9 g and 22.4 g. There were significant differences between the experiments, the results obtained at 72 days of economic yield under glass 17.6 g and in the open field with 9.5 g of economic yields. The agricultural yields obtained in this experiment of 4.9 and 6.4 $kg.m^{-2}$ are acceptable for peppers in the region's environmental conditions.

2.8. Nutritional balance

It is one of the most complete vegetables in terms of nutritional substances. In addition to Vitamin C, they are rich in carotene (Provitamin A), which is found in greater quantities in red fruits, with an oxidised content of between 0.5 and

12.6 mg %. Red fruits contain an abundance of Vitamins B1, B2 and PP. They also contain sugars, fats and proteins, carbohydrates, calcium (CA)phosphorus (P) and iron (Fé). values in their nutrients, depending on the form in which we consume them: dried, green, ripened and fresh.

Nitrogen (N) is one of the constituent elements of proteins, it favours the rapid growth of the meshes and organs of plants, mainly for the formation of leaves and stems, peppers require greater quantities of this element at the time of flowering, if there is a deficiency of nitrogen (N) the upper leaves are small and take on a yellow-green colour and the lower ones fall off very easily, stunting noted in the plants, in addition when they reach flowering they often fall off without ever fertilising.

Excessive nitrogen (N) and phosphorus deficiency can be detrimental, as they lead to disproportionate growth of the plant's vegetative organs, which does not correspond to the development of the roots.
(P) and potassium (K) results in the fruit's flavour being altered and the fruit being less resistant to storage.

Phosphorus (P) takes part in vital combinations that are important for plants (nucleic acids and lecithin). It takes part in the formation of cell nuclei, chromosomes, cell plasma and inzemas; it counteracts the effects of unilateral nitrogen (N) nutrition, an issue that we must take into account when applying nitrogen fertilisations. Phosphorus (P) favours the formation of roots, especially fibrous ones. Plants with good phosphorus nutrition also shorten their vegetative cycle and ensure better formation of the generative organs

If there is a deficiency of this element, the upper section of the leaf shows an anthracite pigmentation and these parts become reddish in colour, and if you look at the root development, you can see that it is stunted. Potassium (K) is very important for the transport of carbohydrates from one organ of the plant to another. Deficiency of this element is reflected in the small leaves, which turn yellow and then burn, and influences fruit formation (María and López, 2020).

MATERIALS AND METHODS

3.1. Location

The experiment was carried out in the province of Cunene in the locality of Oshomocuyo on the property of Joao Kambanda, from March to June 2023, where annual rainfall reaches 4000 mm, and the rainfall for the period is shown in Annex 1, the relative humidity is 80 % and the temperature varies between 27 and 39 °C. The topography the horticultural production area is characterised by slightly flat terrain with no pronounced slopes

3.2. Type of research

The type of research used in this study was experimental and applied, because it involved an experiment in the field using a randomised block design in which the variables were assessed. To this end, four planting distances were evaluated with three replicates of each treatment. The data obtained came from measurements of the plants located in each plot with the aim of obtaining the relationships between them, so that relative conclusions could be reached according to the vegetative cycle of pepper cultivation in field conditions for the region and fulfil the proposed objectives.

3.3. Research methods

The research method uses an experimental design, as the aim is to explain the behaviour of peppers grown at different planting distances and their morphological behaviour through different variables that will be measured, addition to crop yields.

This method consists of deliberately organising conditions according to a prior plan in order to investigate possible cause-effect relationships, exposing one or more empirical groups to the action of an experimental variable and contrasting their results with the same or another comparison group.

3.4. Experiment management

The experiment was carried out in a vegetable production area with the following dimensions: the lengths of the plots were 3.70 metres long and 1.20 metres wide, with a space between plots of 0.70 metres for a total area of 120 $metres^2$.

3.4.1. Research design

The experimental design used for this planting study was a randomised block design, consisting of four treatments with three replications, using 12 plots (Figure 1).

0.40 m *0.20 m T1R1	0.40 m*0.50 m T4R2	0.40 m*0.40 m T3R3
0.40 m*0.30 m T2R1	0.40 m *0.20 m T1R2	0.40 m*0.50 m T4R3
0.40 m*0.40 m T3R1	0.40 m*0.30 m T2R1	0.40 m *0.20 m T1R3
0.40 m*0.50 m T4R1	0.40 m*0.40 m T3R2	0.40 m*0.30 m T2R1

Figure 1. Sketch of the experimental design

3.4.2. Treatments

Four treatments relating to planting distance were used for growing peppers in the open field with the following combinations of distances between rows and between plants.

- T1: 0.40 m * 0.20 m
- T2: 0.40 m * 0.30 m
- T3: 0.40 m * 0.40 m
- T4: 0.40 m * 0.50 m

3.5. Preparing the area for planting

The land was cleared using materials suitable for the soil, using machetes and shovels to remove trunks and roots. With the help of a hoe, the soil was moved to a depth of 0.30 cm, and then each plot was delimited with its different planting distance in the work area.

3.5.1. Planting

The peppers were transplanted after 28 days, taking into account that the seedlings had 4 true leaves, good stem thickness and good health.

3.6. Cultivation work during the pepper growing season

We carried out weevil control once a week throughout the pepper's growing season, using a mallet and hoe to move the soil and remove weevils (Figure 2).

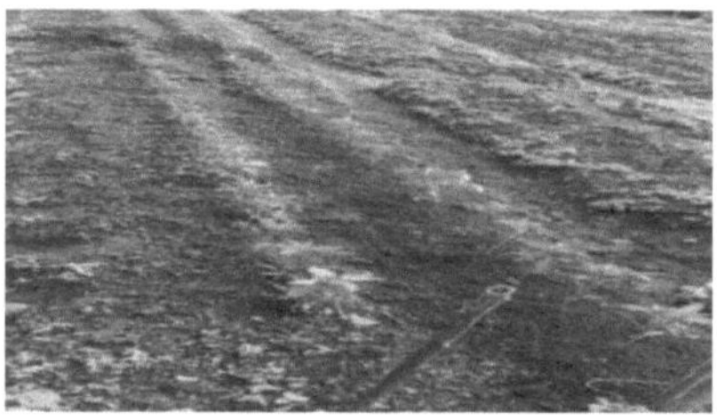

Figure 2. Disease control.

3.6.1. Plant health control

To prevent pests in the experiment, Cypermethrin, an insecticide from the pyrethroid group, was applied at a dose of 10 cm^3/20 litres of water to the foliage of the plant at 5 and 10 days after planting.

3.7. Fertilisation

As a base fertiliser, 1kg of organic fertiliser was applied per plant, which allowed the plant to quickly form roots that would help it feed, grow and bear fruit. In addition, 1.5kg of extra fertiliser was added to each plant 30 days after planting so that the fruit that continues to come off is of good quality and size.

3.8. Measurements and observations made when growing peppers under experimental conditions

Morphological and physiological evaluations were carried out every 10 days until the culmination of the crop's vegetative cycle, selecting 10 plants at random from each replicate.

- **Plant height (cm):** measured from the soil surface to the apical bud using a flexometer (Figure 3) 10 to 70 days after planting in order to determine the growth dynamics of the crop.

Figure 3 Measuring plant height.

- **Number of leaves:** true leaves were counted during the vegetative cycle of 10 pepper plants/replicate (Figure 4).

Figure 4 Number of stems

➢ **Stem diameter (cm):** a caliper was used for this measurement, measuring 1 cm above the ground.

➢ **Leaf length (cm):** measurements were taken from the base to the end of the leaf (apex), using a rule or tape measure (cm).

➢ **Leaf width (cm):** the measurement was taken from the left to the right margin of the leaf in the centre, using a rule or measuring tape (cm).

➢ **Number of flowers and fruit:** the start of flowering and fruiting of the peppers was assessed when the plants reached 15 % their flowering and fruiting stage, and the plants were considered to be massively fruiting when they reached 75 % fruiting stage (Figure 5).

Figure 5 Fruits per plant

➢ **Harvesting time:** when the fruit had ripened or reached a bright green colour on 60% of the plants, the first harvest was carried out at 70 days.

3.9. Physiological measurements of peppers (Capsicumm annuum L.)

➢ **Accumulation of green mass in the vegetative organs of Capsicumm annuum L. (g):** At 30 and 60 days after planting, a random plant from each treatment was taken and each vegetative organ weighed.

➢ **Accumulation of dry matter in the vegetative organs of Capsicumm annuum L. (g):** At 30 and 60 days, one plant was taken at random per replicate of each treatmentOnce weighed in green matter, it was dehydrated to determine dry matter of each organ such as the root, stem, leaves, flowers and fruit.

➢ **Determination of leaf area (m^2):** At 30 and 60 days, the length and width of all the leaves of the 10 plants per treatment were measured, where the average of both variables was taken, and then both variables were summarised to obtain the leaf area.

➢ **Photosynthetic potential (PF):** Calculated using the two leaf area values measured, using the following formula:

$$PF = \sum \left(\frac{A_f + A_i}{2}\right) \cdot t$$

where:

✓ PF: photosynthetic potential

✓ Af: end area

✓ Ai: initial area

✓ t: time

Photosynthetic potential (PF) expresses the average leaf area of living leaves that has worked over the course of the plant's cycle.

➢ **Clear assimilation rate (NAR):** at 30 and 60 days, the amount of dry mass produced by the leaf area unit in one working day was calculated, expressed as $g.m^{-2}.\ day^{-1}$. The formula was used to calculate this:

$$TAN = \frac{2 \cdot (P_f - P_i)}{(A_f - A_i) \cdot (t_f - t_i)}$$

Where

Pf= initial dry weight per plant (first assessment)

Pi= final dry weight per plant (second assessment)

Ai= initial leaf area per plant

Af= final leaf area per plant

tf= is the end time in emergency days,

ti= is start time on emergency days

➢ **Biological yield (BR):** the dry weight of all organs (leaf, root, stem) was summarised for each treatment at two points in time.

➢ **Economic yield (ER):** determined by the dry matter produced by the leaves of the plant in each treatment.

➢ **Agricultural yield (AR):** at 73 and 60 days in both experiments, agricultural yield was determined ($kg.m^{-2}$), where the total weight of the fruit and the number of fruits per plant were averaged over a square metre.

3.10. Study Factors

3.10.1. Independent variables

Different planting distances were used for both experiments, with each distance having 3 replicates (**T1**) 0.40 * 0.20; (**T2**) 0.40 * 0.30; (**T3**) 0.40 * 0.40 and (**T4**) 0.40 0.50, where
evaluated 10 plants per replicate.

3.10.2. Dependent variables

Evaluation of plant height (cm), stem diameter (mm), number of leaves, length and width of leaves, number of flowers and fruit, fresh and dry matter.

3.11. Statistical analysis

$$y_{ij} = \mu + \tau_i + R_j + \varepsilon_{ij}$$

The statistical model corresponding to the exposed design and defined by the following mathematical model was used:

Where:

➢ y_{ij}: variables measured in the experiment

➢ μ: general mean for all observations

- τ_i :Treatment effect i= 1, 2, 3 and 4,
- R_j: Effect replicates j =1, 2 and 3
- ε_{ij}:Normally distributed random error with half 0 and constant variance

The corresponding ANAVAs were carried out and Tukey's multiple comparison test at 5 % was used if the differences between treatments were significant. To analyse the growth of the species, graphs were made showing the behaviour of the species in the different variables.

RESULTS AND DISCOUNT

4.3. Variation in morphological indicators from different planting distances in pepper cultivation

4.3.1. Growth dynamics of the pepper crop (Capsicum annuum) (var. California Wonder)

Figure 6 shows the growth in height of the peppers grown in the open field under semi-desert conditions from the 10th to the 80th day, showing that the treatment with the best behaviour was T1 0.40 x 0.20 m with an average height of 36.03 cm, which differed significantly ($p \leq 0.05$) from the other treatments studied throughout the growing cycle, followed by treatments T4 and T3 which reached a maximum height of 28.60 cm and 31.85 cm respectively. Treatment T2 had the lowest height 23.23 cm.

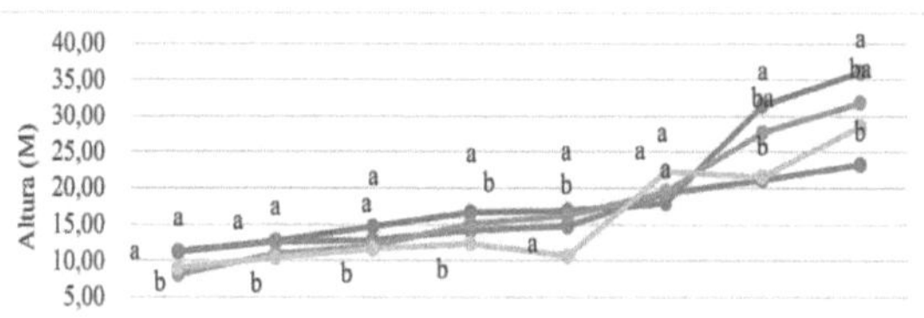

	10 días	20 días	30 días	40 días	50 días	60 días	70 días	80 días
T1	11,30	12,75	14,70	16,68	16,91	17,97	31,48	36,03
T2	11,23	12,65	12,90	14,20	14,80	19,28	21,15	23,23
T3	8,12	10,99	12,08	15,08	16,21	19,52	27,77	31,85
T4	8,97	10,45	11,65	12,38	10,67	22,28	21,57	28,60

Legend: Different letters denote statistical difference for p?0.05 with Tukey's test.

Figure 6. Growth dynamics of different planting distances for growing peppers (Capsicum annuum).

If we compare these results with those obtained under controlled conditions for the same variety of pepper, we realise that the height is lower with a difference of 65.02 cm, which does not correspond to the same treatment which was the one that reported the best results, and which was T2 with a planting distance of 0. 40 m x 0.30 m. These results may be associated with the fact that the pepper was exposed to different abiotic conditions in terms of light, temperature and relative humidity and grew less in height,40 m x 0.30 m. These results may be associated with the fact that the peppers in the open field were exposed to different abiotic conditions in terms of light, temperature and relative humidity, growing less in height, which did not exceed 0.40 cm for all the planting distances studied. Collantes (2015) states that peppers are a small annual shrub between 0.75 and 1.0 metres tall, belonging to the Solanaceae family, with a fragile, erect, green stem and branches that are subdivided into two parts. Alemão et al. (2017) also states that in the open field, pepper plants reached a height of 63 cm after transplanting, which is below the average obtained for peppers in semi-desert conditions.

4.3.2. Stem thickness

As far as this morphological parameter is concerned (Figure 7), it can be seen that up to approximately 25 days there were no significant differences between the planting distances studied, but from 30 to 80 days there were no statistical differences between the treatments studied. The treatments with the best behaviour were T3 and T1 with 0.62 cm and 0.59 cm respectively, with no difference between them, followed by T4 with 0.53 cm and finally T2 with the smallest stem thickness of 0.44 cm. This result indicates that the seedlings that are approximately 5 to 6 cm tall are of higher quality for transplanting because they don't flop over and adapt more easily to the new field conditions. The treatments T3 0.40 m x 0.40 m and T1 0.40 m x 0.20 m are of higher quality, as they are taller and have an adequate thickness, which may be related to the appropriate planting distances for the pepper's vegetative development in the semi-desert region, not favouring the lignification of the tissues, making them more fibrous and therefore not limiting secondary growth in thickness.

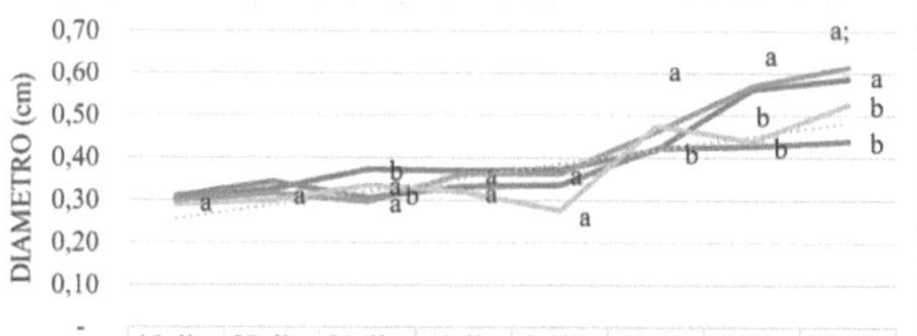

	10 días	20 días	30 días	40 días	50 días	60 días	70 días	80 días
T1	0,30	0,33	0,37	0,37	0,37	0,42	0,56	0,59
T2	0,31	0,34	0,31	0,33	0,34	0,42	0,43	0,44
T3	0,30	0,32	0,30	0,36	0,36	0,46	0,57	0,62
T4	0,29	0,30	0,34	0,32	0,28	0,48	0,44	0,53

Legend: Different letters denote statistical difference for p≤0, 05 with Tukey's test.

Figure 7. Stem diameter at different planting distances pepper cultivation.

These results do not correspond with those of Pérez et al. (2014), who reported that pepper plants obtained a stem diameter of between 1.5 cm and 1.7 cm in a row, which allows for good support of branches and fruit.

4.3.3. Number of leaves

Figure 8 shows the number of leaves the peppers had during their cycle

During the establishment period of the open field crop up to 60 days, the formation of true leaves was slow for all the treatments studied, between 11 and 14 leaves, with no significant difference between them. This was not the case from 60 days until the culmination of its vegetative cycle at 80 days, where leaf growth was rapid, with a significant difference between the treatments ($p \leq 0.05$). The best results corresponded to treatment T1 with 38 leaves, followed by treatments T3 and T4 with 33 and 29 leaves respectively, while the treatment with the lowest number of leaves was T2 with 24 leaves.

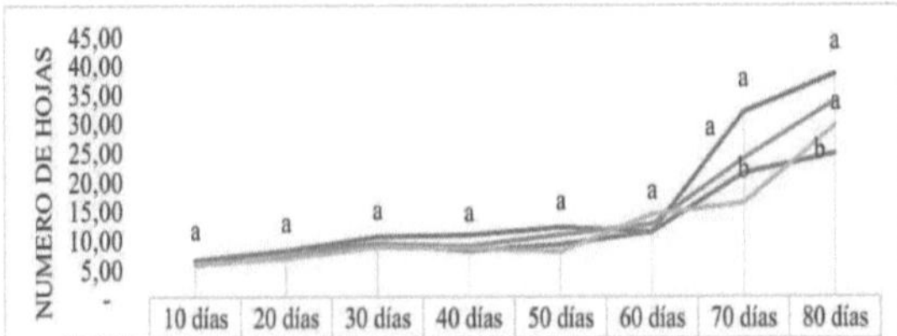

	10 días	20 días	30 días	40 días	50 días	60 días	70 días	80 días
T1	6,33	8,07	10,47	10,77	12,10	11,30	31,77	38,37
T2	6,43	8,10	9,30	8,07	9,13	11,30	21,30	24,73
T3	5,70	7,07	9,27	8,90	10,80	12,67	23,80	33,73
T4	5,87	6,77	8,67	8,53	7,93	14,33	16,20	29,43

Legend: Different letters denote statistical difference for p≤0, 05 with Tukey's test.

Figure 8. Number of leaves at different planting distances in pepper cultivation.

This is an indicator to consider when either growing peppers open field conditions for up to 60 days after planting, or when the crop is adapting to high rainfall conditions, or when the height and number of leaves may be limited, If these two parameters are compared with the results obtained under controlled conditions, it is therefore necessary to improve the soil conditions with the use of organic fertilisers during this period such as the application of organic fertilisers at a rate of 1.5 kg/plant, improving the structure of the soil and boosting the development of peppers. leaves in a row 5 cm to 8 cm long and approximately 3 wide, increasing the number of leaves and preparing the plants for the reproductive phase. Cabalceta and Monge (2017) point out that smaller, more compact plants with adequate leaf cover and good vegetative development for growing peppers in open fields reduce the risk of the fruit being burnt by sun exposure. These results do not correspond to those reported by Alemão et al. (2017), who reported that leaf production under glass and in the open field was 50.3 and 80.3 leaves, with significant differences at 72 days after transplanting, and is an indicator to consider for pepper cultivation due to its direct relationship with photosynthetic capacity.

4.3.4. Number of flowers

With regard to the number of flowers for peppers at four planting distances, it can be seen in (Figure 9) that these began to appear after 25 days, while the number of flowers is reduced between treatments, with very little difference between them up to 60 days. From the 70th and 80th days onwards, there were significant differences between the treatments studied ($p \leq 0.05$), showing that the best treatment was T1 with 14 flowers, followed by treatments T3 and T4 with 13 and 10 flowers respectively. The treatment with the fewest flowers was T2 with 9 flowers per plant.

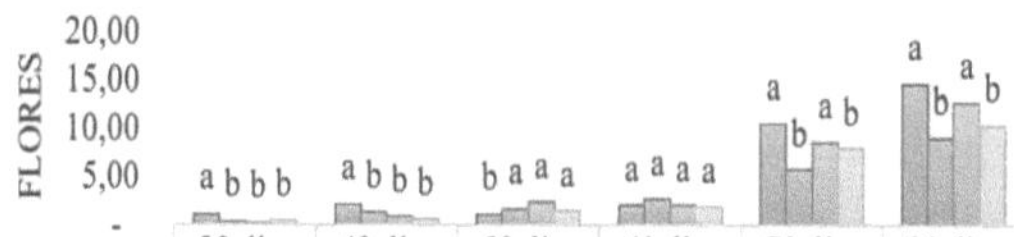

	30 días	40 días	50 días	60 días	70 días	80 días
T1	1,00	2,07	1,00	1,93	10,47	14,57
T2	0,30	1,20	1,60	2,57	5,67	8,93
T3	0,13	0,77	2,27	2,00	8,53	12,63
T4	0,33	0,47	1,37	1,80	7,87	10,27

Legend: Different letters denote statistical difference for $p \leq 0, 05$ with Tukey's test.

Figure 9. Flowering variable for the different pepper planting distances.

This means that the plants with the largest size and number of leaves were found in treatment T1, and these were able to capture a greater amount of light energy, which affected the formation of flowers and therefore the reproductive stage, adapting to the abiotic conditions of the study area in Ondjiva province. When comparing these results with those obtained in the nursery, they are lower. Santos, Segura and Ñústez (2010), in the cultivation of peppers using biofertilisers and suitable planting milestones, the photosynthetic surface and the dry mass content were adequate, forming greater numbers of flowers, which means that the plants are able to capture a greater amount of light energy in relation to their size, if these are compared with treatments without using them.

4.3.5. Number of fruits per plant

The number of fruits per plant for the four pepper planting distances (Figure 10), reports that the first fruits appear from 50 days after transplanting, observing that the four planting distances have a similar behaviour in this morphological parameter at 50 to 60 days, detecting no statistical differences $p \leq 0.05$ with an average fruit number of two fruits per plant. At 70 and 80 days, there were significant differences for this indicator, with the best behaviour being T1 0.40 cm x 0.20 cm with 2.43 fruits, followed by treatments T3 and T4 with 1.04 and 1.57 fruits per plant. The treatment with the lowest number of fruits was T2 with 0.90 fruits, a behaviour associated with greater height during its biological cycle.

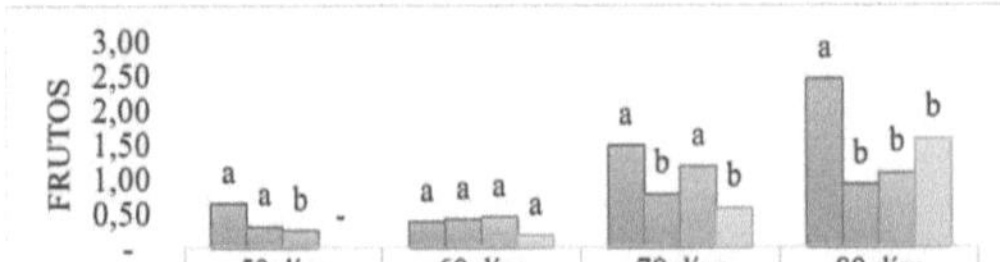

	50 días	60 días	70 días	80 días
T1	0,63	0,37	1,47	2,43
T2	0,30	0,40	0,77	0,90
T3	0,24	0,43	1,17	1,07
T4	-	0,17	0,57	1,57

Legend: Different letters denote statistical difference for p≤0, 05 with Tukey's test.

Figure 10. Fruits per plant in the pepper cultivation area of the experiment.

The result in terms of the number of fruits/plant is low because the plants have a large number of flowers and very small fruits or fruits in the formation stages, which may be due to the fact that the plants are in the open, possibly because they are exposed to high rainfall conditions during their biological cycle in the reproductive phase, which is frequent in the semi-desert region, combined with the effect of nocturnal fog and high temperatures on sunny days, which is very frequent in this region, means that they maintain flowering and fructification. constant, extending the life cycle of the plants by taking advantage of the planting milestones and sunlight. Pérez et al. (2014) indicate that, for peppers grown in the open field, applying efficient microorganisms such as biofertilisers was able to develop a greater leaf area per surface area of soil and a greater number of fruits than other treatments, which translates into better use of the optimum established planting framework and sunlight.

4.4. Effect of four planting distances for open field pepper cultivation in semi-desert conditions on physiological indicators and yield

4.4.1. Accumulation of fresh and dry pepper mass

Table 1 shows the fresh and dry weight values of the peppers at two points in time from transplanting at 30 and 60 days. In relation to the green and dry weight of the root, stem and leaves, it can be seen that the treatment with the best behaviour in terms of this physiological indicator was T1 for fresh and dry weight at both points in time, which may be due to greater vegetative growth in terms of height, number of leaves and stem thickness.

Table 1. Fresh and dry weight values of each vegetative organ of the pepper plant (g).

FRESH MASA 30 days				DRY MASA 30 days			FRESH MASA 60 days			DRY MASA 60 days		
Treatment.	R	H	T	R	H	T	R	H	T	R	H	T
T1	0.6	1.1	0.84	0.1	0.45	0.09	8.8	30.3	29.2	1.3	3.8	4.1
T2	0.6	0.9	0.45	0.1	0.19	0.12	6.2	24.8	24.4	0.9	3.1	3.6
T3	0.9	1.0	0.50	0.2	0.19	0.09	8.2	25.7	22.9	1.1	3.2	3.2
T4	0.6	1.3	0.45	0.1	0.29	0.11	6.6	26.1	25.3	0.9	3.3	3.6

Leyenda: Treatments (TR), Root (R), Hojas (H), Tallo (T)

It is also shown that at 60 days the dry weight of all the organs of the pepper plant was higher for T1 under field conditions, with a total weight of 9.1 g and a stem weight for this variety of 4.1 g, which is lower than those reported by Vale (2010) who obtained values ranging from 11.5 to 14.4 g at 84 days and those reported by Alemão et al., (2017) who obtained a weight of 9.5 g for the same variety under semi-desert conditions. As for the leaves, the behaviour is similar for treatment T1 with a weight of 3.8 g, which is lower than that reported by Vale (2010) who obtained weights of 20.3 - 25.06 g at 84 days, as well as for the leaves. roots the same treatment showed higher percentages of dry mass with statistical differences to other treatments.

4.4.2. Variation in leaf area index of treatments at 30 and 60 days

For the leaf area index (Table 2), when growing peppers under field conditions at four planting distances, it can be seen that there are no significant differences between the treatments studied. It can be seen that the treatment with the largest leaf area corresponds to T1 0.40 m x 0.20 m with a leaf area of 0.11 m^2. Followed by treatments T2, T3 and T4 with the same values of 0.09 m^2 of leaf area.

Table 2. Leaf area index at 30 and 60 days after planting for pepper crops.

FOLIAGE AREA (m2)		
Treatments	30 days	60 days
T1	0.07 a	0.11 a
T2	0.07 a	0.09 a
T3	0.07 a	0.09 a
T4	0.07 a	0.09 a

Legend: Different letters denote statistical difference for p≤0, 05 with Tukey's test.

This result could be due to the fact that the T1 treatment resulted in greater height under field conditions, reflecting a response to the environmental conditions, which are very changeable, and to the different planting distances that were studied. In addition, Barroso (2002) states that leaf area is essential for evaluating indicators of physiological efficiency in the study of the response of varieties and hybrids to different environments or management regimes. García and Tulio (2019), indicate that the leaf area index at 40 days for peppers grown with or using foliar biostimulants (FitoMas - E and Microorganismos Eficiente and Biobras - 16) in field conditions showed a significant difference between them at 60 days, with the treatment with the highest leaf area index or the treatment with Microorganismos Eficiente having 0.11 and 0.21 cm^2, using planting distances of 0.40 m x 0.20 M. These results do not correspond to those obtained by Alemão et al. (2017) for the cultivation of the Nathalie pepper variety under Amazonian conditions in an open field, which was 2.1 m^2.

4.4.3. Photosynthetic potential in the open field (PF)

Table 3 shows that the treatment with the highest photosynthetic potential was T1 with 2.82 $m^2.day^{-1}$ in the open field, followed by other treatments T2 and T3 with the same results of 2.29 $m^2.day^{-1}$ and the treatment with lowest photosynthetic potential was T4 with 2.19 $m^2.day(^{-1)}$.

Table 3. Photosynthetic potential of two treatments

This result could be due to the fact that the T1 treatment resulted in the greatest height in field conditions, reflecting a response to the environmental conditions of how semi-desert is very changeable and to the different planting distances that were studied. In addition, Barroso (2002) states that leaf area is essential for evaluating indicators of physiological efficiency in the study of the response of varieties and hybrids to different environments or management regimes.

García and Tulio (2019), indicate that the leaf area index at 40 days for peppers grown with or using foliar biostimulants (FitoMas - E and Microorganismos Eficiente and Biobras - 16) in field conditions showed a significant difference between them at 60 days, with the treatment with the highest leaf area index or the treatment with Microorganismos Eficiente having 0.11 and 0.21 cm^2, using planting distances of 0.40 m x 0.20 m. These results do not correspond to those obtained by Alemão et al. (2017) for the cultivation of the Nathalie pepper variety under Amazonian conditions in an open field, which was 2.1 m^2.

4.4.3. Photosynthetic potential in the open field (PF)

Table 3 shows that the treatment with the highest photosynthetic potential was T1 with 2.82 $m^2.day^{-1}$ in the open field, followed by other treatments T2 and T3 with the same results of 2.29 $m^2.day^{-1}$ and the treatment with lowest photosynthetic potential was T4 with 2.19 $m^2.day(^{-1)}$.

Table 3. Photosynthetic potential of two treatments.

Photosynthetic Potential	
Treatments	m2.day-1
T1	2.82
T2	2.29
T3	2.29
T4	2.19

If we compare this indicator with the results obtained under controlled conditions (greenhouse) for pepper cultivation, we can see that it is higher, which could be related to the fact that the plants grew less, although the number of leaves is higher than in the greenhouse, another determining factor could be atmospheric and soil temperature, since the plants where the temperature was higher (greenhouse), In addition, Kasperbauer (2000) suggests that the light reflected from the surface of the plastic can alter the light environment sufficiently to modify the photosynthetic rate or morphogenetic development of plant species.

4.4.4. Sharp Assimilation Rate (SUT)

Table 4 shows that the treatments with the best behaviour were T2 0.70 $g.m^2.day^{-1}$ and T3 0.64 $g.m^2.day^{-1}$, which produced the greatest dry mass per m^2 of leaf area in one day. The treatments with the lowest dry mass were T1 and T4.

Table 4: Assimilation rate of the four treatments studied

Neta Assimilation Rate	
Treatments	(g.m2.day(-)(1))
T1	0.50
T2	0.70
T3	0.64
T4	0.51

These results could be playing a game of dice because treatment T1, despite performing better in terms of leaf area, height and the other morphological indicators assessed, had a lower TOU of 0.50 g.m^2.This may be due to the fact that most of the photosynthates produced were distributed to the reproductive parts of the plant, reducing or contributing to vegetative development. Konings (1989) mentions that the variation in TON is generally determined by the variation in the rate of photosynthesis per unit leaf area and the low values of leaf area represent a low photosynthetic capacity of the plant, consequently influencing a low accumulation of dry mass, in addition (Morais et al., 2011) expose that this index registers the speed of photosynthesis in a clear lapse of time, and it is necessary to know the change in two values of leaf area and dry weight of the plant, expressed in g.m^2.día^{-1}

4.4.5. Biological, economic and agricultural performance of pepper cultivation under experimental conditions

The treatment with the highest biological yield at 30 days (Table 5) was T4 with 0.68 g/plant, followed by T1 with 0.50 g/plant. The treatments with the lowest biological yield were T2 and T3 with 0.43 g/plant, showing no significant difference between them in terms of this indicator, nor for the economic yield.

It was found that the treatment with the highest economic yield was T4 with 3.82 g, the result of a greater accumulation of leaf dry mass at 73 days, followed by treatments T1 and T3 with 3.26 g and 3.16 g, respectively. The treatment with the lowest economic yield was T2 with 3.07 g of leaf dry mass.

Table 5: Biological, economic and agricultural yields for growing peppers.

Biological Yield (g/plant)			Economic Income (g)		Agricultural Income (kg/m2)
Treatments	30 days	60 days	30 days	60 days	73 days
T1	0.50 a	7.9 a	0.29 a	3.26 a	4.82 a
T2	0.43 a	7.7 a	0.19 a	3.07 a	3.90 a
T3	0.43 a	7.7 a	0.19 a	3.16 a	3.20 a
T4	0.68 a	9.4 a	0.45 a	3.82 a	3.90 a

Legend: Different letters denote statistical difference for $p \leq 0, 05$ with Tukey's test.

As for crop yield, the treatment with the highest yield was T1 with a planting distance of 0.40 m x 0.20 m; with a total of approximately 3 fruits per plant and a yield of 4.82 kg/m^2, followed by treatments T2 and T4 with 3.90 kg/m^2. These results could be due to the environmental conditions in relation to rainfall, with very frequent rain during the day and the effect of night fog, an adequate planting framework which maintained constant flowering and fructification, extending the useful life cycle of the plants, these results do not coincide with those reported by Alemão et al, (2017) for the same variety of pepper in similar conditions where the yields obtained were 6.4 kg/m^2, and Rodríguez et al. (2014) for the production of peppers in the open field with different hybrids, the highest yield values were 1.15 kg.plant^{-1}, with significant differences between them of 0.77 kg.plant^{-1}.

CONCLUSIONS

1) When growing peppers of the California Wonder variety in field conditions, in terms of morphological indicators under semi-desert conditions, the best treatment was T2 with a planting distance of 0.40 m x 0.30 m, with an average height of 101.32 cm, a stem diameter of 0.9 cm and 37 fruits per plant.

2) The best physiological index under field conditions corresponds to a planting distance of 0.40 m x 0.30 m, a higher green and dry weight gain for roots, stems and leaves compared to other treatments, a leaf area index of 0.34 m^2, a photosynthetic potential of 1.16 m^2/day, a TON of 2.52 and 1.53 g.m^2.day^{-1} and a crop yield of 4.13 kg/$m^{(2)}$.

3) When growing the California Wonder pepper variety in the open field, in terms of morphological indicators in semi-deserts, the best treatment was T1 with a planting distance of 0.40 m x 0.20 m, with an average height of 36.03 cm, a stem diameter of 0.62 cm and 2.43 fruits per plant.

4) The best physiological indicators in the open field correspond to T1 with a planting distance of 0.40 m x 0.30 m with a greater gain in green and dry weight in the root, stem and leaves compared to the other treatments, a leaf area index of 0.11 m^2, a photosynthetic potential of 2.82 m^2/day^{-1}, a TON of 0.70 to 0.50 g.m^2.day^{-1} and a crop yield of 4.82 kg/$m^{(2)}$.

RECOMMENDATIONS

5) Increasing the sowing of peppers in the region

6) Study the California Wonder variety in depth in the open field, as it has a higher yield with a planting distance of 0.40 m x 0.20 m

7) Repeat the experiment using the same planting distances in other areas of the province.

8) Sowing peppers of the California Wonder variety in the open field in the different locations in Cunene province with a suitable sowing distance of 0.40 m x 0.30 m

BIBLIOGRAPHY

Agarwal, A., Gupta, S., & Ahmed, Z. (2007). Inflence of plant densities on productivity of bell pepper (Capsicum annuum) under greenhouse in high altitude cold desert of ladakh.Acta Horticulturae, 756, 309–314. https://doi.org/10.17660/ActaHortic.2007.756.32.

AGRIPAC, (2000). Tomato production under cover. Quito-Ecuador. P.14-18

AGRIPAC, (2012). Yields de híbridos disponibles en; htt //www. linkagro.com/inicio/34-importadora.../329-semillasdepimento 2008. Retrieved 29 March 2019.

Agropecuarios (2014). Cultivation of peppers. España disponible en: http://agropecuarios.net. Alemán, R.; Bravo, C.; Mercedes, O. (2014). Possibilities of producing vegetables in the Amazon region of Ecuador, province of Pastaza.

Alemán, R.; Domínguez, J.; Rodríguez, Y., Soria, S.; Gutierres, R.; Vargas, J. & Alba, L.; (2017). Morphophysiological and production indicators for peppers. Available at: http://scielo.sld.cu/pdf/cag/v45n1/cag02118.pdf

Aminifard, M. H.; Aroiee, H.; Ameri, A. & Fatemi, H. (2012). Effect of plant density and nitrogen fertilizer on growth, yield and fruit quality of sweet pepper (Capsicum annuum). African Journal of Agricultural Research. 7(6): 859-866.

Arias Zambrano, L. J. (2013). Agronomic behaviour of 4 pepper hybrids (Capsicum annuum) in the Luz de América parish, Santo Domingo canton.

Ayala-Tafoya, F., Sánchez-Madrid, R., Partida-Ruvalcaba, I., Yáñez-Juárez, Ruiz-Espinosa, F. H., Velázquez Alcaraz, T., Valenzuela-López, M. y Martín Parra-Delgado, j. 2015. Producción de pimiento morrón con mallas sombra de colores. rev.fitotec.mex., 38 (1): 93 - 99.

Azofeifa A. y M. A. Moreira (2004) Análisis de crecimiento del chile jalapeño (Capsicum annuum cv. Hot), en Alajuela, Costa Rica. Agronomía Costarricense 28:57-67.

Barroso, L.; (2002). Phenology of white albahaca (Ocimum basilicum L.) grown at different harvesting dates. Cultivos Tropicales, 23 (2): 5 - 8.

Berrios, E., Arredondo C., Tjalling H. (2007). "Guía de manejo de nutrición vegetal de especialidad". Peppers. Ediciones Cropkit SQM. Chile.

Borbor Neira, A. and Suárez Suárez, G. (2013). Production of three pepper hybrids (Capsicum annuum) from seeds subjected to imbibition and imbibition plus magnetic field in the Río Verde experimental field, Santa Elena canton (Doctoral dissertation).

Cabalceta, E.; Monge Pérez, J., (2017). Morphological characterisation of 15 pepper genotypes (Capsicum annuum. L) grown under winter in Costa Rica.

Cartagena Pozo, A. S. (2004). Prefeasibility study for the production of green pepper (Capsicum annuum) for export.

Datasheet Version 6 updated (2023). Logistics and industrial centre Qbox-Bod. 25, Copacabana, Antioquia (Colombia) https://www.germisemillas.com octubre del 2023.

Collantes J. (2015). Study of two types of chemical and organic fertilisers two commercial pepper hybrids (Capsicum annuun L) in the upper part of the Guayas River Basin.

Corporación Colombiana de Investigación Agropecuaria (Corpoica) and Gobernación de Antioquia (2014). Pepper production model under protected conditions in the Antioquean East. Retrieved 30 July 2022 from https://repository.agrosavia.co/bitstream/handle/20.500.12324/13748/75921_6584.pdf?sequence=1&isAllowed=y

Domínguez, (2009). Treatise on fertilisation. First edition. Editorial Mundo prensa.Madrid - Spain.

Eco Farmer. 2014. Consulted on 01 April 2019. Available at:www.ecoagricultor.com/el -cultivo-del-pimento/

FAO. 2010. Food and agriculture organisation of the United Nations. Statistics on pepper cultivation. http://faostat.fao.org/site/339/default.aspx

Flexas, J.; Barón, M.; Bota, J.; Ducruet, J. M.; Gallé, A.; Galmés, J.; & Tomàs, M. (2009). Photosynthesis limitations during water stress acclimation and recovery in the drought-adapted Vitis hybrid Richter-110 (V. berlandieri× V. rupestris). Journal of Experimental Botany, 60(8), 2361 -2377.

García y Tulio (2019). Effect of four foliar biostimulants on the physiology and yields of peppers (capsicumm annuum). Revista Infociencia, ISSN 1029- 5186, Vol.23, N.1, enero-abril, 2019, p. 59- 70.

Grijalva-Contreras, R. L.; Macías-Duarte, R. & Robles-Contreras, F. (2008). Productivity and quality of chile bell pepper varieties and densities under wintering conditions in north-west Sonora. Biotecnia (México). 10(3): 3- 10.

Gruda, N.; 2005. Impact of environmental factors on product quality of greenhouse vegetables for fresh consumption. Critical Reviews in Plant Sciences, 24: 227-247.

Jovicich, E.; Cantliffe, D. J. & Hochmuth, G. J. (1999). Plant density and shoot pruning on yield and quality of a summer greenhouse sweet pepper crop in Northcentral Florida. En: K. D. Batal (Ed.). Proceedings 28th National Agricultural Plastics Congress, 19-22 May (pp. 184-190). Tallahassee, Florida, USA.

Jovicich, E.; Cantliffe, D. J. & Stoffella, P. J. (2004). Fruit yield and quality of greenhouse- grown bell pepper as influenced by density, container, and trellis system. Hort Technology. 14(4): 507-513.

Kasperbauer, M. J. 2000. Strawberry yield over red versus black plastic mulch. Crop Sci.40: 171-174.

Konings, H. 1989. Physiological and morphological differences between plants with a high NAR or a high LAR as related to environmental conditions. pp. 101 -123. In: H. Lambers, M. L. Cambridge, H. Konings, and T. L. Pons (eds.). Causes and consequences of variation in growth rate and productivity. SPB Academic Publishing. The Hague, the Netherlands.

Maria Judo, M. López Hernandez (2020). Tabla provisional de la comparación nutritiva de alimentos cubanos, Lab, FIM, de Nutrición, B.S.M.S, L Habana.

Martínez, P. S. (2005). Conjunto Tecnológico para la Producción de (P. S. Martínez, Editor) Retrieved 29 March 2019, from Soil and land preparation: http://www.eea.uprm.edu/sites/default/files/PIMENTO-

Martínez, S. (2011). Efecto de cuatro bioestimulantes en el crecimiento y productividad del cultivo de pimento (Capsicum annuum) Variedad cacique en la zona de Chaltura, Provincia de Imbabura". Pages 22-35

Morales Rosales, J. E., Francomora, O., González Huerta, A. 2011. Snap bean production using sunflowers as living trellises in the central high valleys of México. Cien. Inv. Agr., 38 (1): 53-63.

Morales Rosero, E. J., & Pachacama Cevallos, S. L. (2011). Agronomic evaluation of five sweet pepper hybrids (Capsicum annuum) with three doses of chemical fertilisation, under winter in the parish of Pifo.

Moreno Pérez, E.; Mora Aguilar, R.; Sánchez del Castillo, F.; García Pérez, V. (2017) Phenology and yield of hybrids of Pimento morrón (Capsicum annuum) grown in hydroponics.

Moreno, E. C.; Mora, R.; Sánchez, F. & García-Pérez, V. (2011). Phenology and yield of pepper hybrids (Capsicum annuum) grown in hydroponics. Revista Chapingo Serie Horticultura. 17(special issue 2): 5-18.

Moreno, V.; Ribas, F. E.; Cabello, M. J. (2004). Extract from an article in the journal Agricultura. Pág. 476-480

Nogueira, I.; María Teresa Gomes; López, R.; da Silva, A. and Maria do Rosário L. (2007). Producto e cualidades de frutos de híbridos de pimentão (Capsicum annuum) en ambiente protegido en Manaus-AM". Acta Amazónica. vol. 37(4): p. 491 - 496.

Orellana C. and León E. (2011). Evaluating the production of 3 varieties of pepper (Capsicum annuum) grown hydroponically under winter in La Molina nutrient solution.

Pérez Toapanta, M. F. (2014). Evaluation of three substrates and four doses of biostimulants for the production of ornamental peppers (Capsicum annuum) under winter conditions. Quito, Pichincha.

Pérez, Y. L., & Barrera, Y. P. (2014). Efecto de la aplicación del bioestimulantes fitomase en tres etapas de desarrollo del cultivo del pimento Eumed.net. Retrieved from: http://www.eumed.net/rev/delos/20/cultivo-frijol.html

Raven, P. H., Evert, R. F., & Eichhorn, S. E. (1992). Plant biology (Vol. 2). Reverté.

Reséndiz-Melgar, R. C.; Moreno-Pérez, E. C.; Sánchez-Del Castillo, F.; RodríguezPérez,J. E. & PeñaLomelí, A. (2010). Pimento morrón varieties managed with early harvesting at two population densities. Revista Chapingo Serie Horticultura. 16(3): 223-229.

Reyes Pérez, J.; Luna Murillo, R.; Reyes Bermeo, M.; Zambrano Burgos, D.; Vázquez Morán, F., (2017). Fertilisation with organic fertilisers in peppers (Capsicum annuum) and its impact on yield and its components. Pág. 90-93

Ríos A. (2012). del cultivo, E. D. L. P., pimento, d., de Fertilización, Provincia, Pichincha, D., Endara, G., & Ecuador, G. (2012). Bolívar State University, Faculty of Agricultural Sciences, Natural Resources and the Environment, School of Agronomic Engineering.

Ríos, J. P. (2011). The country will export fresh sweet chile to the United States. La Nación (Costa Rica). 30 August. Obtenido desde: http://www.nacion.com/economia/Paisexportara- chile-fresco-Unidos_0_1216878334.html.

Rodríguez Llanes, Y.; Tomás, L.; Depestre, M.; Palloix, A. (2014), Behaviour in the open field of new f1 hybrids and varieties of pepper (Capsicum annuum) Multidrug resistant to viruses. Pág. 52-57.

Román Esquivel, C. (2017). Fertiliser technical sheet. Agausa Agrobiologuicos

Seifi, S.; Nemati, S. H.; Shoor, M. & Abedi, B. (2012). The effect of plant density and shoot pruning on growth and yield of two greenhouse bell pepper cultivars. Journal of Science and Technology of Greenhouse Culture. 3(11): 77-

83.
Serrano, C. Z. (1996). Twenty winter vegetable crops. Manuales técnicos de Sevilla, España. pp. 433-487. ISBN: 84-605-4596-2.
Suquilanda, M. 2002. Organic pepper production in the Northern and Central Sierra of Ecuador. Quito. P. 18.
Taiz, L., & Zeiger, E. (2010). Plant physiology 5th Ed. Sunderland, MA: Sinauer Associates, 782 p.
Valle, M.J.C. (2010). Biomasa accumulation, growth and nutrient extraction (Capsicum annuum). Horticulture Institute. Universidad Autónoma Chapingo. Chapingo, Mexico. 90 p.
Valles Rigió, G., Lugo González, J., Rodríguez, Z. F., & Díaz, L. T. (2009). Effect of the substrate and the sowing distance between plants on the growth of pepper plants (Capsicum annuum) in a hydroponic system without cover. Revista de la Facultad de Agronomía, 26(2), 159-178.
Villavicencio A. Vásquez, W. (2008). Technical Crop Guide. Instituto Nacional Autónomo de Investigaciones Agropecuarias Manual Nº 73. Sheet 1 and 2. Revised on 30/03/2019.

ANNEXES

Annex #1. Climatic variables

MAPA DE TEMPERATURAS MÁXIMA E MINIMA, HUMIDADE RELATIVA E PRECIPITAÇÃO REFERENTE AO I TRIMESTRE DE 2023 – DADOS CLIMATICOS

Dias	Tª Máxima °c			Tª Mínima °c			Humidade Relativa %			Precipitação mm		
	Jan	Fev	Marc	Jan	Fev	Marc	Jan	Fev	Marc	Jan	Fev	Marc
01	30,3	29,2	33,1	19,3	20,0	19,7	70	61	51			
02	30,4	30,9	29,8	21,0	21,5	19,1	61	51	66			
03	23,7	30,1	32,8	19,8	21,3	19,8	67	67	44			
04	20,1	32,1	33,0	18,7	20,8	18,0	98	48	20			
05	29,8	31,4	34,0	19,0	22,2	18,4	40	30	22			
06	31,1	30,9	35,9	19,6	20,8	18,8	29	59	21			
07	33,7	26,7	35,9	20,8	20,9	21,4	31	58	35			
08	32,3	28,6	31,0	22,9	21,7	21,5	27	65	50			
09	31,3	31,7	34,7	21,2	19,4	22,6	30	43	30			
10	33,2	33,1	34,4	23,1	18,8	22,4	44	46	40			
11	28,9	33,1	32,0	21,2	20,0	20,8	75	31	78			
12	31,6	33,8	32,8	21,6	20,3	22,6	67	31	68			
13	30,7	35,1	34,8	19,9	21,7	21,0	62	35	40			
14	31,1	35,6	35,0	18,0	23,4	21,2	68	31	39			
15	31,9	35,2	34,3	18,0	21,6	21,1	50	22	33			
16	29,6	33,3	33,4	20,7	18,2	20,2	71	27	47			
17	26,4	32,9	31,7	20,0	16,9	17,7	67	31	40			
18	25,7	31,9	33,5	19,0	13,7	20,0	85	23	50			
19	25,1	33,0	33,3	18,1	15,0	19,4	96	22	46			
20	22,3	33,1	33,9	20,6	15,4	19,3	90	24	49			
21	27,2	34,2	32,6	18,6	14,8	20,2	78	22	55			
22	24,6	35,4	32,4	17,2	17,4	21,3	71	26	47			
23	27,4	34,0	29,8	19,4	21,3	18,8	79	39	72			
24	28,4	32,9	32,0	18,5	20,8	20,5	79	48	53			
25	30,2	34,5	30,8	20,6	20,5	18,5	68	49	46			
26	31,0	34,3	25,6	19,0	20,2	21,2	71	46	84			
27	31,4	32,5	30,6	20,2	21,6	18,3	70	64	55			
28	30,1	27,9	28,4	21,1	19,7	18,6	68	-	77			
29	30,3		31,0	20,1		20,3	59		66			
30	27,8		31,8	17,9		20,6	77		63			
31	27,4			19,6		18,9	76		62			
Total	901,6	906,3	1006,6	548,8	548,8	622,2		1099,0	1565,0			
Média	28.6	30,2	32,5	19,7	19,6	20,1	65	78	51			

MAPA DE TEMPERATURAS MÁXIMA E MINIMA, HUMIDADE RELATIVA E PRECIPITAÇÃO REFERENTE AO II TRIMESTRE DE 2023 – DADOS CLIMATICOS

Dias	Tª Máxima °c			Tª Mínima °c			Humidade Relativa %			Precipitação mm		
	Abril	Maio	Junho	Abril	Maio	Junho	Abril	Maio	Junho	Abril	Maio	Junho
01	31,1	33,0	31,5	20,2	16,7	13,8	61	30	25			
02	32,0	32,5	32,1	19,4	15,1	15,5	59	27	29			
03	32,2	33,0	31,3	20,6	14,4	13,0	50	30	27			
04	31,8	32,5	32,7	20,6	14,4	12,4	55	31	22			
05	31,2	32,3	31,8	18,8	16,5	11,4	43	30	21			
06	32,6	32,6	31,8	17,8	16,5	11,4	54	25	13			
07	33,8	33,1	31,6	18,0	15,2	13,4	45	31	17			
08	33,4	32,8	30,9	19,6	17,8	12,6	39	35	21			
09	32,6	32,8	30,4	20,3	16,3	13,4	42	40	19			
10	33,6	33,5	31,2	21,3	16,3	13,4	46	24	20			
11	33,9	32,8	30,7	19,5	18,3	11,4	34	26	20			
12	33,4	31,7	27,2	19,5	15,1	11,8	40	30	22			
13	34,4	31,8	29,2	18,0	15,9	7,8	39	37	17			
14	33,7	31,7	29,4	19,8	14,2	9,2	35	28	24			
15	33,9	31,5	29,4	18,5	15,0	9,3	37	21	21			
16	33,4	30,2	30,1	16,6	19,9	7,7	33	34	15			
17	32,6	30,3	30,1	17,7	13,4	7,8	37	30	15			
18	32,7	30,5	30,0	16,7	11,2	9,2	45	24	18			
19	33,4	31,0	29,9	18,7	11,3	7,3	46	25	22			
20	31,6	32,3	30,0	20,2	12,3	8,1	45	18	22			
21	20,9	32,4	29,7	17,9	13,3	10,8	75	20	22			
22	29,2	33,1	29,6	18,0	15,6	10,8	72	25	20			
23	32,5	32,4	29,3	19,4	14,5	9,0	45	20	21			
24	32,7	32,4	28,3	20,6	16,4	11,2	46	28	22			
25	32,6	31,1	28,5	20,8	15,2	11,0	56	23	27			
26	33,3	31,7	28,5	19,1	13,4	9,5	48	21	34			
27	33,6		23,0	18,2	15,1	9,4	39	23	52			
28	33,5	31,7	22,4	18,2		3,3	28		32			
29	33,7	31,9	27,1	16,4	15,9	3,6	35	20	21			
30	33,4	29,8	28,2	16,0	12,6	6,8	33	23	27			
31		30,7			11,5			22				
Total	973,6	959,1	885,9	566,4	449,3	305,3	1362	801,0	661,0			
Média	32,4	31,9	29,5	18,9	14,9	10,1	45	27	22			

MIX
Papier aus verantwortungsvollen Quellen
Paper from responsible sources
FSC® C105338

Printed by Books on Demand GmbH, Norderstedt / Germany